SPACE EXPLORERS

OUR SOLAR SYSTEM

SPACE EXPLORERS

OUR SOLAR SYSTEM

A thrilling voyage through our cosmic backyard

DR VIENNA TRAN · illustrated by DONOUGH O'MALLEY

little smith

THE SOLAR SYSTEM

Our Solar System is a collection of planets, asteroids and other objects all moving around the Sun.

It formed about 4.5 billion years ago when a large cloud of gas and dust collapsed, forming the Sun and the planets we know today. The rocky planets are Mercury, Venus, Earth and Mars, while the gas giants are Jupiter, Saturn, Uranus and Neptune. The asteroid belt sits between Mars and Jupiter and consists of millions of rocky, metallic and icy objects over 1 km (.62 mi) in diameter.

DID YOU KNOW?

You can remember the planets with the phrase: 'My Very Educated Mother Just Served Us Nachos'. The first letter from each word corresponds to the first letter of each planet in order.

Many of the planets have their own moons that orbit them. Other objects in the Solar System include asteroids, comets, meteoroids and dwarf planets like Pluto. Since 1994, we have discovered thousands of other planetary systems in the Milky Way, all with a group of planets bound to a host star by gravity. We call our planetary system the 'Solar System' because 'solis' is the Latin word for the Sun.

Humans have been fascinated by the Solar System since prehistoric times. We first observed the planets with the naked eye, then with telescopes. All the planets in the Solar System have been visited by spacecraft launched from Earth. These missions have taken photographs of the planets and gathered useful scientific data.

THE SUN

Our Sun is a large, hot ball of hydrogen and helium at the centre of the Solar System.

The Sun measures 150 million km (93.2 million mi) across and weighs the same as 330 060 Earths. Because of its mass, it exerts a strong gravitational force that holds the entire Solar System together, from the smallest dust particles to the largest planets.

In about 5 billion years, the Sun will run out of energy and expand into a giant red star so large that it will engulf Mercury, Venus and perhaps Earth. It will expel most of its outer material and spend the rest of its life as a dense white dwarf star.

The Sun is only one of countless stars in the universe. The closest star to the Sun is Proxima Centauri, about 4 light years away from Earth. That means it takes light four years to travel from Proxima Centauri to us.

The Sun is responsible for the weather on Earth and throughout the Solar System. It goes through an 11-year cycle of high and low activity. During a solar maximum, the Sun throws out much more energy and matter, affecting equipment such as satellites, radio communications and even power grids. Scientists use telescopes and probes to predict these solar events and study how the Sun changes the space environment.

DID YOU KNOW?

Most of the mass of the Solar System is contained within the Sun – 99.86%, to be exact!

MERCURY

Mercury is the smallest planet in our Solar System and the closest to the Sun.

Mercury has no atmosphere, which means it gets as cold as -180°C (-292°F) and as hot as 430°C (806 °F): the largest daily temperature variations in the whole Solar System. It has a large metallic core, which takes up most of Mercury's radius and could be partly molten.

If you went to Mercury, you would notice that it is grey and dusty like the Moon. In fact, the planet is only slightly larger than the Moon. With no atmosphere to protect it, Mercury's surface has been bombarded with meteoroids and asteroids over time, creating craters as wide as 1.5 km (.93 mi). Because of its temperature extremes and high solar radiation levels, Mercury would be a dangerous place for life to exist.

The spacecraft MESSENGER visited Mercury in 2011 and orbited the planet over 4000 times, taking over 100 000 images and mapping the entire surface.

DID YOU KNOW?

Since it spins so slowly, one day on Mercury is equivalent to 176 Earth days! On top of that, the years are shorter than the days, with three Mercury days equal to two Mercury years.

VENUS

Venus is known as Earth's twin. It is similar in size and is Earth's closest neighbour.

The planet was named after the Roman goddess of love. Easily seen on a clear morning or evening, Venus is known as the most beautiful star in the sky. But don't be fooled! Venus has a dense, toxic atmosphere made of carbon dioxide, nitrogen and sulphuric acid. These gases create an extreme greenhouse effect, trapping the Sun's heat on the surface.

Because of its thick atmosphere, Venus is the hottest planet in the Solar System, even though it is not the closest to the Sun. Temperatures can reach a scorching 464°C (867°F): hot enough to melt lead. On top of that, the atmospheric pressure at the surface of Venus is almost 100 times the pressure at sea level on Earth.

Because of its harsh environment, exploring Venus can be difficult. The sulphuric acid clouds are opaque, and we cannot see the surface with optical telescopes from Earth. A lander called Venera 7 was sent to the Venusian surface in 1970 and survived. Two decades later, the Magellan spacecraft visited Venus to map the surface using radar, which could see through the thick clouds.

DID YOU KNOW?

If you could stand on Venus, you would witness a backward sunrise: the Sun rises in the west and sets in the east.

EARTH

Earth is our home planet and the third planet from the Sun.

At its centre, the Earth has an iron core that generates a magnetic field. This magnetosphere protects the Earth from the solar wind, a stream of charged particles released from the Sun's atmosphere. The magnetosphere also protects us from cosmic rays that come from deep space.

Earth's surface is made of large tectonic plates that are always moving. Over millions of years, they have collided and moved past each other to make mountains, valleys and many more geological wonders.

Earth sits in the 'Goldilocks Zone': an orbit that is not too hot and not too cold but just right for water to be in its liquid state. For life as we know it to exist, liquid water must be present. Water gives structure to cells (the building blocks of life) and allows important chemical reactions to occur. More than 70% of our planet's surface is covered in water, making it unique in the Solar System and allowing life to flourish.

The first person to leave Earth and go to space was Russian cosmonaut Yuri Gagarin in 1961. Since then, many people have been to space to live, work and explore.

DID YOU KNOW?

After seeing Earth from space, many astronauts gain a stronger sense of connection to others and the Earth itself. This is called the 'overview effect'.

THE MOON

The Moon is a rocky body that orbits the Earth once every 28 days.

The first person to set foot on the Moon was Neil Armstrong in July 1969, followed by his crewmate Edwin 'Buzz' Aldrin. Many other vehicles, with and without astronauts, have also landed on the surface, leaving behind hundreds of artefacts like country flags, tools and even golf balls!

The Moon has no protective atmosphere surrounding it, which means there is no wind to wipe away astronauts' footprints. The lunar surface is constantly bombarded by meteorites, leaving holes or craters behind.

Our Moon most likely formed out of a collision between Earth and another small planet billions of years ago. Debris broke away and collected in an orbit around the Earth, forming the Moon we know today.

Scientists have found ice in craters at the north and south lunar poles, where it is cold and dark. Other areas closer to the equator can get very hot, with temperatures reaching 114°C (237°F).

One day you may be lucky enough to view a lunar eclipse: when the Sun, Earth and Moon are aligned. Light from the Sun passes through the Earth's atmosphere, casting a brilliant red glow on the Moon's surface.

DID YOU KNOW?

If you stood on the Moon, you would weigh about one-sixth as much as you do on Earth.

DID YOU KNOW?

Sunsets on Mars are blue! This is because the dust particles in the atmosphere scatter light from the setting Sun, allowing blue light through more easily.

MARS

Mars is our planetary neighbour and the fourth planet from the Sun.

Mars was named after the Roman god of war. It is well known as the 'red planet' for its red soil, rich in iron oxide. You can spot Mars as a bright point of reddish light in the night sky.

Mars has many similarities to Earth. A Martian day is about 24.6 hours. Mars is also tilted at 25 degrees, giving the planet four seasons. However, its atmosphere is 100 times thinner than Earth's, so it's not as good at warming the planet or protecting it from meteoroids. Despite the thin atmosphere, the wind on Mars can be strong, reaching 100 km/h (62 mph). Mars is famous for its intense dust storms that can cover most of the planet and block out the sunlight.

To call Mars home, humans would need to construct a new world. A Martian habitat should provide oxygen, temperature control, protection from Martian winds, and protection from harmful solar radiation.

Since the 1970s, spacecraft, including rovers, landers, satellites and a helicopter, have been sent to Mars. We have learned that our neighbouring planet was once warmer and covered in flowing rivers of water.

MOONS OF MARS

Mars has two small moons that orbit closely: Phobos and Deimos. In Greek mythology, Phobos ('fear') and Deimos ('dread') were the sons of Ares, the god of war.

The two moons were discovered by an American astronomer called Asaph Hall in 1877. He waited for several days in stormy weather, finally catching a glimpse of the moons through his telescope.

Unlike Earth's spherical Moon, Phobos and Deimos are irregular in shape. Their origin is still unknown, but most scientists believe that they were once asteroids in the asteroid belt, captured by Mars' gravity.

Phobos has a radius of 11 km (6.8 mi) and completes a full orbit in just 11 hours. Its surface is pockmarked with craters; the most obvious one is the 9-km-wide (5.6-mi-wide) Stickney Crater, which looks like a large dent on one side. Its orbit around Mars is synchronous, which means that the same face of Phobos is always facing the planet.

Deimos is slightly smaller with a radius of 6 km (3.7 mi) and it orbits farther away. Its surface appears much smoother than Phobos's because many of its craters are filled with dust or regolith. It would look like a star from Mars' surface.

DID YOU KNOW?

If you stood on Deimos and jumped straight up, you would escape its gravity and launch into space.

DID YOU KNOW?

Some of the wackiest names given to asteroids include James Bond, Tomhanks, Mr Spock and Jabberwock.

ASTEROID BELT

The asteroid belt is a region of asteroids that lies between the orbits of Mars and Jupiter.

It contains over 1 million asteroids ranging from 10 m (33 ft) to 530 km (329 mi) in diameter, all orbiting the Sun.

In its very early stages, the Solar System was made of smaller particles of gas and dust. Slowly, the planets came together under the force of their own gravity. However, Jupiter's massive gravity has made it impossible for a full-sized planet to form between Mars and Jupiter. To this day, the gravity fields of Mars and Jupiter still exert their force on the asteroids, pushing and flinging them in all directions.

Asteroids have even collided with Earth, changing the course of our geological and evolutional history. All the precious metals found on the Earth's surface or within the crust, like gold and platinum, were deposited by asteroids after the Earth's crust solidified.

Today, scientists are keeping a close eye on near-Earth objects (NEOs), such as asteroids, that pose a risk of impacting our planet. If an NEO is too close for comfort, we could send a spacecraft to crash into it, deflecting it away from disaster.

CERES

Ceres is a dwarf planet residing in the asteroid belt.

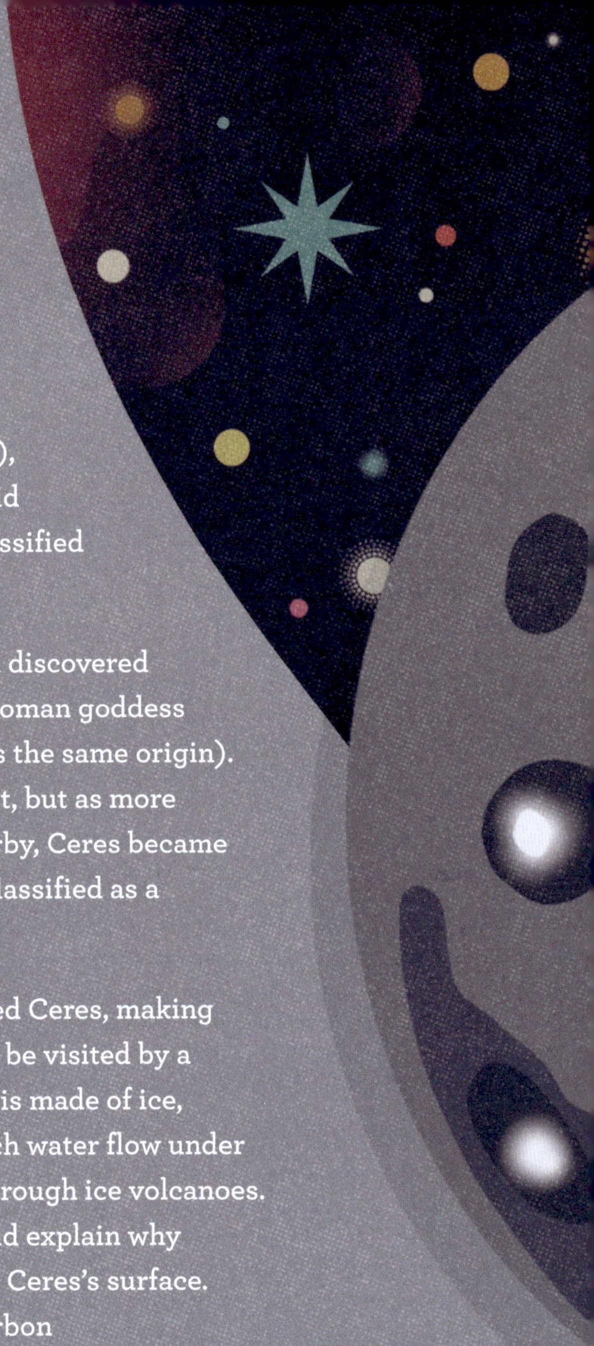

With a diameter of 476 km (296 mi), it is the largest object in the asteroid belt and the only one that is not classified as an asteroid.

Italian astronomer Giuseppe Piazzi discovered Ceres in 1801, naming it after the Roman goddess of agriculture (the word 'cereal' has the same origin). It was originally labelled as a planet, but as more rocky objects were discovered nearby, Ceres became classified as an asteroid. It was reclassified as a dwarf planet in 2006.

In 2015, the Dawn spacecraft orbited Ceres, making Ceres the first dwarf planet to ever be visited by a spacecraft. Dawn found that Ceres is made of ice, carbon and clay. Streams of salt-rich water flow under the surface, sometimes erupting through ice volcanoes. This exciting volcanic activity could explain why water vapour has been detected on Ceres's surface. The large volumes of water and carbon make Ceres a possible habitat for life.

DID YOU KNOW?

Ceres comprises a quarter of the asteroid belt's total mass.

JUPITER

Jupiter is the largest planet in the Solar System and the first of the gas giants.

Its mass is more than twice that of all the other planets in the Solar System combined. It is named after the king of the Ancient Roman gods.

Most of Jupiter is made of hydrogen and helium. Close to the inner rocky core, the pressure and temperature are so high that the hydrogen gas becomes a liquid ocean.

If you look at Jupiter through a telescope on a clear night, you may notice its reddish-orange stripes. These stripes are large clouds of ammonia and water vapour. With no solid ground to stop storms, the weather can be violent on Jupiter, with constant lightning and winds reaching 360 km/h (224 mph).

The Great Red Spot is the most famous feature of Jupiter. It is a 400-year-old giant storm twice as wide as the Earth, with winds reaching 650 km/h (404 mph). Its red colour is believed to be from a chemical reaction involving ammonia and sunlight. Much of the Great Red Spot remains a mystery because Jupiter's thick clouds make it difficult to see.

DID YOU KNOW?

If Jupiter were only 75 times more massive, it could have become a star. In other words, our Solar System may have had two Suns!

MOONS OF JUPITER

CALLISTO

Jupiter's four largest moons are Io, Europa, Ganymede and Callisto.

These moons are also known as the Galilean moons because they were discovered by Italian astronomer Galileo Galilei in 1610. Jupiter has at least 95 known moons, resembling a mini solar system.

Named after Greek mythological figures, each moon has its own unique characteristics. Io is covered in sulphur, giving it a distinct yellow colour. It is constantly being pulled between the gravity of Jupiter and the other Galilean moons. This movement creates friction in its core, making it the most volcanically active body in the Solar System.

Europa's surface is mostly water, ice and rock. Clay-like minerals have also been found, and an ocean of water may lie underneath. These features make Europa one of the most likely places in the Solar System for life to exist outside of Earth.

Ganymede is the largest of the four moons, even larger than Mercury. It is the only moon in the Solar System with its own magnetic field, thanks to its iron-rich liquid core.

Callisto is the 'odd moon out': it is the only Galilean moon without distinct rocky layers, and its orbit is not in sync with the other three moons.

DID YOU KNOW?

If the moons of Jupiter were further away from Jupiter's bright light, they could be seen in the night sky with the naked eye.

DID YOU KNOW?

*If you could put Saturn in water,
it would float like a cork.*

SATURN

Famous for its rings, Saturn is the second-largest planet in the Solar System.

It is made of very similar elements to Jupiter: mostly hydrogen and helium, with a core of metals and rocky elements. It has faint stripes of clouds, which give the planet a yellow-brown colour. On Saturn, the hurricane-force winds are strong enough to squeeze gas into liquid.

The most beautiful feature of Saturn is its large ring system, which can be seen through a telescope from Earth. The rings are made of millions of pieces of rock, ice and dust. Some of these pieces are as small as a grain of rice, and some are as big as a mountain. Originally, they came from objects like comets and asteroids, which were pulled into Saturn's gravity field and broken into pieces. Along with the ring system, Saturn has 83 known moons.

Perhaps the most bizarre feature of the planet's appearance is the hexagon-shaped cloud pattern surrounding its north pole. It measures about 30 000 km (18,641 mi) wide. A large storm swirls in the centre, with winds reaching 320 km/h (199 mph).

The Cassini spacecraft gave us a great deal of information about the ringed planet. It orbited Saturn for 13 years. When it ran out of its fuel in 2017, it was sent straight down into the planet to protect Saturn's moons.

MOONS OF SATURN

Saturn has at least 274 known moons and moonlets (smaller moons) in its orbit, ranging from the size of a football field to the size of Mercury.

Titan, Saturn's largest moon, has many unique features. So far, it is the only place in the Solar System outside of Earth where we have found liquid rivers, lakes and seas on its surface – methane and ethane, to be exact. But to find liquid water, you would simply have to look under the thick crust of water ice. Together, these materials may harbour a new form of life.

The Huygens space probe safely parachuted down to the surface of Titan in 2005, collecting data about Titan's atmosphere and imaging parts of the surface in great detail.

Enceladus is another fascinating moon. It snows on Enceladus, but not as we know it. Icy sprays of liquid burst through cracks in the crust, which have been detected by the Cassini spacecraft. This spray is made of water vapour, methane, carbon dioxide and other gases: some of the important elements for life. Some of the spray becomes part of Saturn's ring system, but most of it falls back down like snow, giving Enceladus its bright white appearance and its cold temperature of -200°C (-328°F).

DID YOU KNOW?

Titan could become more habitable in a few billion years from now. When the Sun becomes a red giant, it could warm Titan enough to support liquid water on the surface.

DID YOU KNOW?

There may be diamonds on Uranus. Science suggests that there could be an ocean of metallic liquid carbon close to the core. The pressure high enough to turn the liquid into solid 'diamond-bergs'.

URANUS

Uranus is one of the ice giants of the Solar System, with Neptune being the other. It was the first planet to be found with the help of a telescope.

Uranus has 28 known moons and 13 narrow rings. It rotates at an angle of nearly 90 degrees from the plane of its orbit, resembling a ball rolling around the Sun. A large Earth-sized object probably crashed into Uranus about 4 billion years ago, knocking it onto its side and causing its bizarre orbit.

More than 80% of Uranus is made of dense icy materials including water, methane and ammonia. The atmosphere is mostly hydrogen and helium with traces of methane gas. This methane is what gives the planet its light blue-green colour.

Because Uranus is so far away, not many spacecraft have visited. In 1986, NASA's Voyager 2 studied the planet for 6 hours after having travelled 3 billion km (1.86 billion mi) through the Solar System.

Other than spacecraft, we can use optical telescopes to study planets. The Perth Observatory in Australia played a role in discovering Uranus' rings when the planet passed in front of a star in 1977.

MOONS OF URANUS

The Uranian moons are thought to have formed from the birth of the Solar System.

MIRANDA

Five main moons dominate the space around Uranus, with many smaller moons orbiting closely between them.

Miranda is named after a character in Shakespeare's *The Tempest*. It is the smallest of the five moons and has many canyons up to 12 times as deep as the Grand Canyon.

Ariel is the brightest Uranian moon, perhaps because of the water ice on its surface. Its name comes from the air spirit in a poem by Alexander Pope. Umbriel was named after a gnome from the same poem. It is the darkest moon, and its quirkiest feature is a bright ring-shaped area on one side, nicknamed the 'fluorescent Cheerio'.

Titania is the largest moon, spanning 1,578 km (980 mi). It is made of frozen water and carbon dioxide. Like Miranda, it has large canyons all over its surface. Oberon is a dark red, heavily cratered moon with a similar composition to Titania. Both Titania and Oberon are fairy characters in Shakespeare's *A Midsummer Night's Dream*.

DID YOU KNOW?

Finding moons is not easy. Sir William Herschel, who discovered Titania and Oberon, was certain that he had observed another four moons orbiting Uranus. He was later proven wrong.

DID YOU KNOW?

Although Neptune is 17 times the mass of Earth, its surface gravity is not much stronger. If you were 60 kg (132 lb) on Earth, you would only weigh 8 kg (17 lb) heavier on Neptune.

NEPTUNE

Neptune is the furthest planet from the Sun.

It is named after King Neptune, the Roman god of the sea. It is the only planet whose existence was predicted by mathematics based on the movement of its neighbour Uranus.

It takes 4 hours for the Sun's light to travel to Neptune. Because of its distance from the Sun, it is the only planet not visible to the naked eye. It is also the coldest planet in the Solar System, reaching a chilly minimum temperature of -221°C (-365°F). If that wasn't harsh enough, Neptune has the strongest winds in the Solar System, reaching speeds of 2000 km/h (1200 mph): faster than the speed of sound.

Uranus and Neptune are made of very similar elements. For a long time, scientists wondered why they were different shades of blue. Today, we know that both planets have a methane haze in their atmospheres. Neptune's winds are turbulent enough to thin out this haze, allowing more of the blue colour through.

Voyager 2 visited Neptune in 1989, finding features like large storms, wispy white clouds made of ammonia, and faint rings.

MOONS OF NEPTUNE

Neptune has 16 known moons, all named after sea characters in Greek mythology.

Triton, named after the Greek sea god, is by far the largest Neptunian moon. It is one of the few moons in the Solar System that are geologically active, with the others being Io, Europa, Enceladus and Titan.

Triton is dotted with volcanoes, and its crust is made of frozen nitrogen and water ice. Active springs, also known as geysers, shoot jets of nitrogen gas as far as 8 km (5 mi) out of the surface. These geysers are responsible for creating Triton's very thin atmosphere.

Interestingly, it orbits in the opposite direction to Neptune's rotation. This is unusual for a moon and suggests that Triton may have been a dwarf planet, captured from the Kuiper belt by Neptune's gravity (see page 41).

Over a short time from 1977 to 1989, Triton's colour was seen to change from a marbly red to a paler hue. This is likely due to nitrogen frost, which has gradually covered the red methane ice.

DID YOU KNOW?

Triton is slowly spiralling towards Neptune. One day, it will either collide with Neptune or be torn apart by gravity, forming a new ring around its planet.

DID YOU KNOW?

Within the Kuiper belt, some objects gravitate so closely that they make contact with each other, forming a peanut-like shape called a contact binary.

KUIPER BELT

The Kuiper belt is a colossal disc containing millions of icy objects orbiting the Sun, that lies just past Neptune.

It was named after the astronomer Gerard Kuiper who speculated in 1951 that there must be more objects beyond Pluto.

The Kuiper belt was likely formed from leftover material from planet formation. Most Kuiper belt objects have been unchanged for more than 4.5 billion years, making them the oldest materials in the Solar System.

At a glance, the Kuiper belt resembles the asteroid belt, only colder, larger and more massive in total. Its main region starts near Neptune and spans about 3 billion km (1.86 billion mi). The other part of the Kuiper belt is the scattered disc, which covers a larger area. Objects in the scattered disc have very eccentric, or elliptical, orbits.

Kuiper belt objects (KBOs) have a wide range of colours, shapes and sizes up to 1000 km (621 mi) wide, which include the dwarf planets Pluto, Makemake, Haumea and Eris. The total mass of everything in the Kuiper belt is no more than 10% of the mass of Earth.

Pluto was the first KBO to be discovered. In 2006, New Horizons was the first spacecraft to fly through the Kuiper belt, visiting Pluto on its way.

PLUTO

Pluto is a dwarf planet made of rock and ice.

It has a very thin atmosphere of nitrogen, methane and carbon monoxide, and its snow is a brilliant red colour. Pluto was discovered in 1930 by 24-year-old Clyde Tombaugh after scientists suspected that there was a planet beyond Neptune affecting its orbit.

Because Pluto is so far away from the Sun, it has a very large orbit: one Plutonian year is equal to 248 Earth years. So, in the year 2178, Pluto will have made its first full orbit since its discovery.

Pluto was originally recognised as a planet, but it was demoted by the International Astronomical Union in 2006. It is now considered a dwarf planet because it has not managed to 'clear the neighbourhood' around its orbit. A full-sized planet has enough gravity to remove other bodies of similar mass in its orbit, whereas Pluto has not done this. Charon, Pluto's largest moon, has a diameter that is over half that of Pluto itself!

The dwarf planet was first visited by the New Horizons spacecraft, which launched in 2006 and flew by Pluto in 2015. On board were the ashes of Clyde Tombaugh, who passed away at the age of 90. The spacecraft captured images of a large glacier shaped like a heart.

DID YOU KNOW?

In 1930, 11-year-old Venetia Burney suggested the name Pluto for a new planet that had been discovered. Her grandfather passed the name on to a group of astronomers who liked it and made it official.

DID YOU KNOW?

Charon does not technically orbit Pluto. The dwarf planet and its moon actually orbit around a common centre of gravity, which is why they are sometimes considered a double dwarf planet system.

MOONS OF PLUTO

Pluto has five moons, all named after mythological figures of the underworld.

CHARON

For such a small rocky body, Pluto has a surprisingly complex moon system. Scientists believe that these moons were created from a collision between Pluto and another Kuiper belt object, flinging out material that eventually became locked into Pluto's orbit.

Charon is by far Pluto's largest moon, measuring half the diameter of Pluto itself. Most of Charon is made of water ice, ammonia crystals and rocky material. There may also be icy geysers and volcanoes on this mysterious moon.

In 1990, the Hubble Space Telescope took the first photo with a high enough resolution to show Pluto and Charon as two separate bodies.

Pluto's other four moons, Styx, Nix, Kerberos and Hydra are irregularly shaped and not spherical like Charon. Their orbits are also much further away from Pluto. They all orbit in nearly perfect circles in the same direction as Pluto's rotation. The smallest of these moons is Styx, only 7 km (4.3 mi) in diameter.

Pluto is not the only dwarf planet with moons. Makemake, Haumea and Eris have their own moons too.

MAKEMAKE

Makemake is a dwarf planet in the Kuiper belt with an interesting origin story.

It was discovered shortly after Easter in 2005 by a team of American astronomers and was at first given the code name 'Easterbunny'. When it was officially announced to the public, the Easter theme stuck. Makemake is the god of fertility in the mythology of the Rapa Nui people, the native people of Easter Island. It is pronounced 'mah-kee-mah-kee'.

Although a year on Makemake is 305 Earth years, its days are similar in length to Earth and Mars: only 22.5 hours. It has a faint 80-km-wide (50-mi-wide) moon named MK 2. Because it is so far away, scientists are yet to learn more about the dwarf planet. We do know that it has a reddish-brown colour, and its surface is covered in frozen snow made of methane and ethane. It usually has no atmosphere, but when it is closest to the Sun, it may develop a thin atmosphere of nitrogen.

Despite being the second brightest object in the Kuiper belt, Makemake was discovered after several other fainter Kuiper belt objects. This was probably because its orbit is tilted at a higher angle where astronomers were less likely to search.

DID YOU KNOW?

Makemake's gravity is very weak. If you stood on Makemake and jumped, you would fly many metres into the air and float slowly back down.

DID YOU KNOW?

Haumea was originally nicknamed Santa because it was found just after Christmas.

HAUMEA

Haumea is a dwarf planet in the Kuiper belt.

It spins very fast compared to other objects, taking the shape of an egg. In fact, a day on Haumea is only 4 hours.

Aside from its peculiar shape, Haumea is also special because it is the first known Kuiper belt object with a ring. In the same way that Uranus's rings were discovered, Haumea's ring was found when it passed in front of a star, blocking some of its light.

This dwarf planet is named after the goddess of fertility in Hawaiian mythology. It has two moons, Namaka and Hi'iaka, named after Haumea's daughters. They were probably formed when a large object collided with Haumea, throwing out large pieces of rock that became the two moons.

No spacecraft has visited Haumea, so we rely on ground-based telescopes to study it. Originally it was thought to be made completely of rock. We now know that its surface is covered in bright crystalline water ice.

The discovery of Haumea is a topic of debate. An American team of astronomers were the first to view it from the Mauna Kea Observatory in Hawaii in 2004. However, a team from Spain is officially credited with the discovery because they were the first to announce it publicly.

ERIS

Eris is the heaviest dwarf planet in the Solar System. It lies in the scattered disc of the Kuiper belt.

Eris doesn't mind straying from the other dwarf planets; its orbit is much more elliptical and higher in angle, so it goes through some extreme seasons. When it is further from the Sun, the methane in its atmosphere condenses and falls like snow. As it comes closer to the Sun, the atmosphere warms up again.

The discovery of Eris in 2003 holds a special place in astronomy's history. It was the first Pluto-sized object to be observed outside of the Solar System. If Pluto was a planet, then faraway Eris had to be a planet too. So, scientists had to redefine what it means to be a planet. Pluto no longer fit the description, so it was 'kicked out' of its category, joining Eris as a dwarf planet.

Eris is named after the Greek goddess of discord and trouble. Its moon is named Dysnomia, after the daughter of Eris. No other names could be more suitable, given how much trouble Eris caused for Pluto!

It is difficult to learn about Eris because, like the other dwarf planets, it is so small and lies billions of kilometres away. Observations from a ground telescope in Hawaii reveal that it is probably made of rock, covered in a smooth layer of methane.

DID YOU KNOW?

Eris is almost a perfect mirror, reflecting 96% of the light that shines on it.

OORT CLOUD

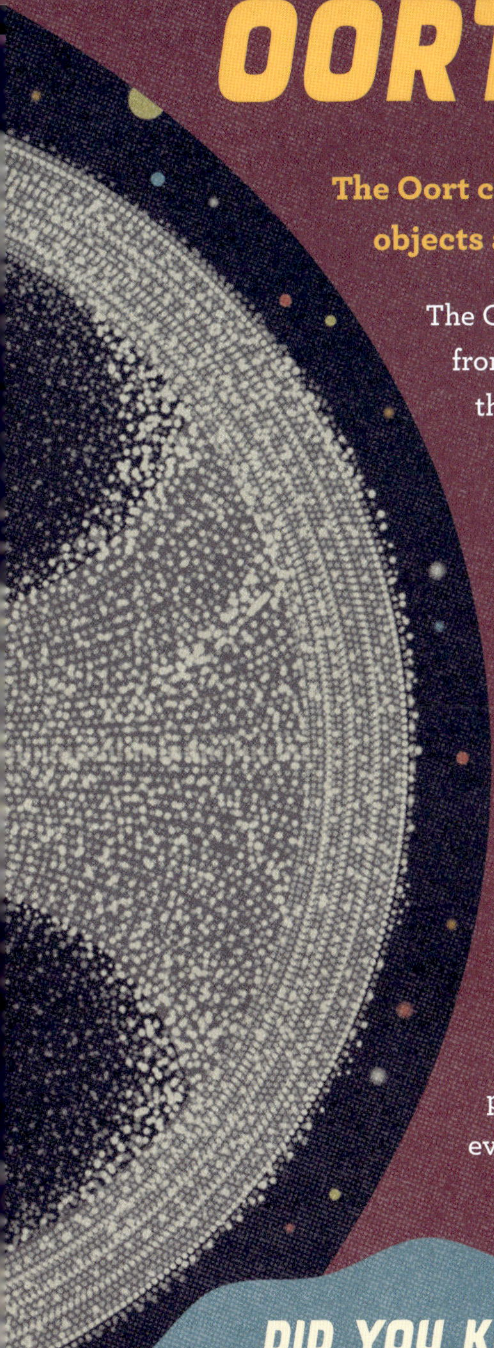

The Oort cloud is a sphere of trillions of icy objects at the edge of the Solar System.

The Oort cloud is unimaginably large. Light from the Sun takes a year and a half to reach the outer edge. It envelops the planets, asteroids and dwarf planets like a bubble.

It is perhaps the most mysterious part of the Solar System because nobody has ever seen it. Dutch astronomer Jan Oort was the first person to come up with the idea. He noticed that many comets passing through the Solar System had very long orbits, so they had to be coming from somewhere further out than the Kuiper belt. This is how the concept of the Oort cloud was created.

If it is real, then it was formed soon after the birth of the Solar System when particles of gas, dust and ice were scattered everywhere. The smaller ice chunks were gradually flung away into the outer Solar System by the larger planets, forming the theoretical cloud.

DID YOU KNOW?

Voyager 1 was launched into deep space in 1977. Even while travelling at 17 km (10.5 mi) per second, it would take 300 years to reach the Oort cloud and 30 000 years to pass through it.

COMETS

A comet is a ball of ice and dust whizzing around the Sun in an elliptical orbit.

When a comet gets closer to the Sun, it warms up and releases gases, forming a miniature atmosphere. This atmosphere forms a beautiful bluish-green 'tail' that can be seen with the naked eye from Earth. The name 'comet' is related to the Greek word for 'long hair'.

There are two types of comets. Short-period comets take less than 200 years to make a full orbit. At their furthest point, they enter gas giant territory.

Long-period comets can take millions of years to complete an orbit. Many of them are thought to exist in the Oort cloud since their orbits are so large. Comet McNaught is one example. When it was visible to us in 2007, it was bright enough to be seen in the daytime sky. Other long-period comets have not passed through the Solar System since the dawn of our species.

The first mission to land on a comet was the European Space Agency's Rosetta, which landed on the comet 67P/Churyumov-Gerasimenko, halfway between the orbits of Mars and Jupiter. Its lander, Philae, sent detailed photos back to Earth.

DID YOU KNOW?

Sometimes, comets can get 'lost': they disappear from where we'd expect to find them. The comet may have been destroyed or shifted around by the gravity of large planets like Jupiter.

ASTEROIDS & METEORS

In our Solar System, there are countless small, rocky objects that orbit the Sun. They have different names depending on their size and location.

Asteroids are leftover fragments of rock from when the Solar System first formed. They are not big enough to be planets. Scientists estimate that there are over a million asteroids orbiting the Sun, with most of them sitting in the asteroid belt.

When a space rock is smaller than 1 metre (3.2 ft) , it is called a meteoroid. Most of them reside in the asteroid belt and are made of nickel and iron. If one of these rocks weighs less than a gram (.04 oz), it is called a micrometeoroid or cosmic dust.

When a meteoroid enters Earth's atmosphere, it is called a meteor. As it falls at a high speed, it creates friction with the atmosphere, becoming a bright fireball or shooting star. If you go stargazing on a dark night, you are bound to see one.

If a meteoroid survives its perilous journey through the atmosphere and hits the ground, it is called a meteorite. Many meteorites have been found in deserts and people's gardens.

DID YOU KNOW?

Occasionally, the Earth will pass through a large collection of meteoroids, creating a spectacular show of shooting stars called a meteor shower.

SATELLITES

Artificial satellites are made by humans and are launched into orbit around Earth and other planets.

Satellites in low-Earth orbit (LEO) can take high-quality images of Earth and provide internet coverage with minimal signal delay. Even though they are in space, LEO satellites still collide with air particles in the atmosphere and can de-orbit after several years.

Many medium-Earth orbit (MEO) satellites perform navigation and location tracking. Geostationary satellites (GEO) can observe one place on the Earth continuously and are great for telecommunications and weather observation.

One famous LEO satellite is the International Space Station (ISS). It is operated by a team of five space agencies from 15 different countries. The ISS is the largest space-based, human-made object in the Solar System, spanning the size of a soccer field. Since November 2000, astronauts have been living on the ISS continuously, conducting experiments and maintaining the station.

The first satellite, Sputnik, was launched in 1957. Today there are over eight thousand satellites in orbit, but only about five thousand of them are active. They can break apart during launch or orbit, creating space 'junk' or debris.

DID YOU KNOW?

If you counted all the satellites and debris in orbit that measure 1 cm (.4 in) or more, you would find about 1 million of them.

DID YOU KNOW?

The Venetia Burney Student Dust Counter (VBSDC) is the first planetary science instrument to be built by students. It was sent aboard the New Horizons spacecraft to gather data about interplanetary dust.

INTERPLANETARY DUST

The space between planets and other objects is not completely empty. The interplanetary dust cloud is the cosmic dust that exists throughout the whole Solar System.

If you have ever seen sunlight shining through your window, illuminating the dust particles that float in mid-air, you can imagine what interplanetary dust might look like.

When you look at the night sky, you would first notice the Moon and the stars. The rest of the sky might look black to the naked eye. However, if you used a specially designed telescope, you would see that the sky is full of infrared radiation (heat). In the Solar System, it is these dust particles that scatter light and emit these tiny amounts of heat.

Interplanetary dust comes from all over the Solar System, including collisions between rocky bodies, the outgassing from comets, and perhaps even the Martian dust storms.

The dust doesn't hang around for long. Most interplanetary dust is new, unlike the rocky objects in the asteroid or Kuiper belts, which are remnants from the Solar System's birth.

DID YOU KNOW?

In the far future, the Milky Way and the Andromeda Galaxy may be pulled together so closely that they become one big galaxy.

MILKY WAY GALAXY

The Solar System is one of many planetary systems in the galaxy called the Milky Way.

When you look at the night sky, you will notice that many of the stars are concentrated in a faint white band, like milk that has been spilled across the sky. This is where the Milky Way gets its name. In fact, the word 'galaxy' comes from the Ancient Greek word for milk.

A galaxy is a huge system of stars, planets, dust and other matter, all bound together by gravity. There are an estimated 2 trillion galaxies in the universe, and the Milky Way is just one of them.

The galaxy spans nearly 100 000 light years across. It is classified as a spiral-type galaxy because it has two main spiral 'arms' and many small ones. The Solar System lies somewhere on the smaller Orion-Cygnus Arm. The arms swirl out from a central bulge 8000 light years thick. A dense group of stars and a supermassive black hole reside in this Galactic Centre.

The Milky Way is locked in a gravitational 'dance' with its neighbour, the Andromeda Galaxy. They are the two largest galaxies in the Local Group, a collection of at least 80 galaxies bound together by gravity.

little smith

Published in 2026 by Little Smith,
an imprint of Smith Street Books
Naarm (Melbourne) | Australia
smithstreetbooks.com

Distributed outside of ANZ, North & Latin America by
Thames & Hudson Ltd., 6-24 Britannia Street, London, WC1X 9JD
thamesandhudson.com

EU Authorised Representative: Interart S.A.R.L.
19 rue Charles Auray, 93500 Pantin, Paris, France
productsafety@thameshudson.co.uk; www.interart.fr

ISBN: 978-1-9235-0327-4

Smith Street Books respectfully acknowledges the Wurundjeri People of the Kulin Nation, who are the Traditional Owners of the land on which we work, and we pay our respects to their Elders past and present.

Publisher: Hannah Koelmeyer
Project editor: Lucy Grant
Design: Sarah Mitchell
Layout: Nikola Roberts
Proofreader: Helen Koehne
Pre-press: Megan Ellis
Production manager: Aisling Coughlan

Printed & bound in China by C&C Offset Printing Co., Ltd.

Book 465
10 9 8 7 6 5 4 3 2 1

FSC MIX Paper | Supporting responsible forestry FSC® C008047